YOUR KNOWLEDGE HAS VALUE

- We will publish your bachelor's and master's thesis, essays and papers

- Your own eBook and book - sold worldwide in all relevant shops

- Earn money with each sale

Upload your text at www.GRIN.com and publish for free

Imprint:

Copyright © 2018 GRIN Verlag
Print and binding: Books on Demand GmbH, Norderstedt Germany
ISBN: 9783668629639

This book at GRIN:

https://www.grin.com/document/389100

Chikaeze Akaolisa

Investigation of the Antibacterial Activity of three types of medicated soaps on Staphylococcus aureus

GRIN Verlag

DEDICATION

To

All those who have always been an encouragement to me to do this work, especially my supervisor, my parents, my lecturers and coursemates.

ACKNOWLEDGEMENTS

I wish to express my sincere gratitudes to my supervisor Prof. Roland Ndip Ndip for giving me the opportunity to work with him, for accepting the topic and for the scientific supervision of this work. I equally thank; Prof. Samuel Wandji and all lecturers and Staff of Department of Microbiology and Parasitology, Faculty of Science, University of Buea (UB) for the skills they have imparted in me for the past three academic years, permitting me to do this work. I also wish to thank Mr. Mbaabe Felix for technical assistance.

I will be forever grateful to my parents, Mr. Akaolisa Joel and Mrs. Akaolisa Maryann for all they had to endure to provide the finances for my studies, as well as for all the pieces of advice and immense contribution they chipped in to steer my imagination in the right direction. A million thanks to my siblings, Laura, Verine, MacDonald and Treasure as well as to every other member of the family and also friends like Pelagie, Julius, Luke and Ewane who put in one thing or another to assist me in this quest.

And above all, I give all the glory to God Almighty for seeing me through.

ABSTRACT

Following invasion of the human body by bacteria, most bacteria produce substances that are toxic to the human body. In the fight against these infections, the production of antibacterial substances (synthetic compounds of different forms) has been a major step in solving this problem. In this study, three medicated soaps: Dettol, Pharmapur and Tetmosol were investigated for their antibacterial activity at different concentrations against *Staphylococcus aureus* using the disc diffusion method. Saline was used as a negative control and a readymade antibiotic impregnated disc (Ciprofloxacin) was used as a positive control. Bacterial identification was by standard microbiological techniques which included: Colonial examination, Gram staining and biochemical testing. Dettol medicated soap had the highest antibacterial activity (26mm at 1/10 soap dilution) while Tetmosol showed the least antibacterial activity (6mm at $1/10^3$ soap dilution). *S. aureus* was found on 4 of the 6 samples that were analysed giving a prevalence of 66.67%. All the medicated soaps showed antibacterial activity which depended on the concentration (dilution) of soap sample on *S. aureus*; hence, the use of antibacterial soaps is recommended as a means of reducing risks of transmission and infection by bacteria.

TABLE OF CONTENTS   **Page**

LIST OF TABLES AND FIGURES   **Page**

CHAPTER ONE

1.0   INTRODUCTION AND LITERATURE REVIEW

1.1 INTRODUCTION

The human skin makes contact with thousands of microbes that are pathogenic to man. When proper hygiene is not practiced, these microbes replicate, produce toxic substances and eventually cause disease. Some of these microbes cause diseases on the skin while others become pathogenic only when they migrate to mucous membranes and other vital organs.

Staphylococcus aureus has been known to be one of the many dangerous and versatile pathogens with high death rates in humans as a consequence of community-acquired and hospital-acquired infections. It has shown little change in mortality rates and a steadily increasing morbidity. *S. aureus* is a member of the Micrococcaceae family; a group of pyogenic cocci known to cause various suppurative or pus forming diseases in humans and other animals. It is a non-motile and non-spore forming facultative anaerobic organism that grows by aerobic respiration or fermentation. On microscopical examination, the organisms appear as Gram-positive cocci in clusters. The distinction of *S. aureus* from other staphylococcal species is very important in clinical microbiology laboratories because this pathogen serves as an important source of nosocomial infections; this is done on the basis of the yellow pigmentation of colonies and positive results of coagulase, mannitol fermentation and deoxyribonuclease test (Lowy, 1998; Murray *et al*, 2003). *S. aureus'* ability to synthesize the enzyme coagulase makes it the most pathogenic of the Staphylococci. Prevention and treatment of infections caused by this pathogen is very important to public health especially with the emergence of multi drug resistant strains which has led to the search for new antibiotics or the modification of existing ones (Lyon and Skurray, 1987).

Soaps play an important role in the prevention and treatment of infections because they have the ability to remove and kill pathogens and this ability is well exploited in skin hygiene (Riaz *et al*, 2009). Skin hygiene, particularly of hands, is considered one of the primary mechanisms to reduce risk of transmission of infectious agents by both the contact and faecal-oral routes. The washing of hands with soaps and water is a routine practice which was established many generations before now as a means to ensure personal hygiene and over the decades, bathing, scrubbing and washing traditions have become established within the health care setting (Larson, 1999; Bhat *et al*, 2011). The importance of hand washing with soap is more crucial when it is associated to health care workers because of possible cross contamination of bacteria and other pathogens that may be pathogenic or opportunistic (Richards *et al*, 1999). Soaps, detergents and hand sanitizers used for hygiene and sanitation are expected to effectively remove infectious agents. It is for this purpose that medicated soaps which possess chemical agents/compounds that are antimicrobial are produced. These soaps produced by different manufacturers have different antibacterial strengths which is due to difference in the concentration and/or antimicrobial content of the chemicals/compounds found in them. The analysis of the antimicrobial activity of these medicated soaps is therefore necessary as a major step to maintaining proper hygiene in our communities.

1.2 LITERATURE REVIEW

1.2.1 The skin and the normal skin microbiota

The $1.8m^2$ human skin has folds, invaginations and specialized niches. This ecosystem inhabits diverse groups of microorganism providing them with shelter and nutrients (Grice and serge, 2011). It is part of the human innate immune system and its primary role is to serve as a physical barrier, protecting our bodies from potential assault by foreign organisms or toxic substances. It is the first line of defense against invasion by infectious agents. The epithelial surfaces of the skin are very impermeable to most infectious agents and desquamation of the skin epithelium also help to remove bacteria and other infectious agents that have adhered to epithelial surfaces thus protecting the human body. However, when the skin is breached, disease can occur due to infection. The skin is also an interface with the outer environment and, as such, is colonized by a diverse collection of microorganisms which may be symbiotes, commensals or parasites. These include; bacteria, fungi and viruses as well as mites. Many of these microorganisms are harmless (symbiotes and commensals) and the symbiotes may in some cases provide vital functions that the human genome has not evolved. Symbiotic microorganisms occupy a wide range of skin niches and protect against invasion by more pathogenic or harmful organisms by competing with them for nutrients and adhesions proteins on the skin or produce substances toxic to them and may also play a role in educating the billions of T cells that are found in the skin, priming them to respond to similarly marked pathogenic cousins (Grice and Serge, 2011). These microorganisms can be categorized either as "resident flora" or contaminants ("transient flora") (Rotter, 1966). According to Nobel and Somerville (1974), the normal flora ("resident flora") of the skin composed primarily of Gram-positive cocci and diphtheroids which may represent a selective barrier against proliferation of potentially pathogenic organisms. In some individuals, small numbers of Gram-negative organisms or yeast may also comprise this normal flora, but their proliferation may be influenced by the normal ecological balance (Aly and Maibach, 1976).

1.2.2 Burden, prevalence and Epidemiology of *Staphylococcus aureus* infections

The diversity of bacteria is seen in their presence and wide distribution all over the surface of the earth. They are found in soil, water, sewage, standing water as free-living organisms and can become pathogenic upon contact with the human body. (Bhat *et al*, 2011; Johnson *et al*, 2002). Infection by a bacterium does not necessarily result in a disease state. Disease occurs when (1) The bolus of infection is high, (2) when the immune system is compromised, (3) when the virulence of the invading organism is great. Transient bacteria are known to be the cause of most bacterial infections. Bacterial infectious diseases represent an important cause of mortality and morbidity worldwide. The rates of diseases resulting in deaths is even worse in developing countries where sanitary conditions are poor, health care services are limited and worse-still few people can afford the health care services presented to them.

Staphylococcus aureus bacteraemia (SAB) represents a significant burden on health care systems. According to an analysis of a large US database, SAB was associated with a longer median duration of hospital stay, higher median total treatment cost, and greater risk of mortality, compared with bacteraemia caused by any other pathogen (Shorr *et al*; 2006). In U.S hospitals in the national nosocomial infections surveillance system, *S. aureus* accounted for up to 13% of isolates recovered from patients with nosocomial infections from 1979 through 1995 and the percentage has increased in recent years. Community-acquired infections with *S. aureus* are also common (Eiff *et al*, 2001). *S. aureus* is the most frequent bacteria pathogen among clinical isolates from hospital in-patients in the United States and is the second most prevalent bacterial pathogen among clinical isolates from out-patients. According to the SENTRY Antimicrobial surveillance program, which examined more than 81,000 isolates during the period of 1997-2002, *S. aureus* was the most common cause of nosocomial bacteraemia in North America (prevalence, 26.0%) and Latin America (prevalence, 21.6%) and the second most common cause of bacteraemia in Europe (prevalence, 19.5%). The incidence of nosocomial MRSA infections has also greatly increased in recent years in the Unites states and Europe (Naber, 2009).

1.2.3 Clinical manifestations and pathogenesis of *Staphylococcus aureus*

The versatility of *S. aureus* is seen in a variety of clinical conditions which include: septicaemia, pneumonia, wound sepsis, septic abortion, osteomyelitis, septic arthritis, post-surgical infections and toxic shock syndrome (Nkwelang *et al*, 2009). These results in high rates of morbidity and mortality in humans worldwide (Murray *et al*, 2004).

Approximately 30% of healthy individuals can asymptomatically carry *S. aureus* on mucous membranes for weeks or months but is only transiently carried on the intact skin. Five stages are seen in the pathogenesis of *S. aureus*. These include: (1) colonization, (2) local infection, (3) systemic dissemination and/or sepsis, (4) metastatic infection, and (5) toxinosis. Once in blood, the organism spreads widely to peripheral sites in distant organs and septic shock can occur. Without specific therapy, the mortality rate resulting from spread of infection is high. Specific infections result from haematogenous dissemination, these include: endocarditis, osteomyelitis, renal carbuncle, septic arthritis, or epidural abscess. Even when the organism itself does not invade the bloodstream, specific syndromes such as toxic shock syndrome, scalded skin syndrome and food borne gastroenteritis result from the local or systemic effects of specific toxins (Archer, 1998).

1.2.4 Resistance of *Staphylococcus aureus* to antibacterial agents.

The excessive use of antibiotics has led to resistance. Resistance to antibiotics in most cases, is coded for by gene carried on plasmids, accounting for the rapid spread of resistant bacteria (Harris *et al*, 2002). *S. aureus* like many other human pathogens has evolved resistance to commonly used antimicrobial agents. Multidrug-resistant *S. aureus* strains have been reported all over the world with increasing frequencies. These include isolates that are resistant to methicillin, lincosamides, macrolides, aminoglycosides, fluoroquinolones or a combinations of these antibiotics. Methicillin resistant strains of *S. aureus* (MRSA), have emerged intermediate resistance to glycopeptides which are the main drugs with reliable activity against MRSA (Eiff *et al*, 2001). This is a major problem to the pharmaceutical industry and has led to the search for new antimicrobials or the modification of existing antimicrobials to increase their spectra of activity.

Methicillin resistance is due to the acquisition of a new penicillin-binding protein, PBP2a. This protein has a low affinity for most β-lactam antibiotics and, therefore, mediates cross resistance to all these compounds. This high level of resistance not only impedes successful therapy for infections but also allows the organisms to persist in the hospital, expanding its reservoir. (Archer, 1998).

1.2.5 Prevention, control and treatment of *Staphylococcus aureus* infections.

Person to person contact transmission is a mode of transmission of infections by certain pathogens. Cleaning of hands immediately after using the toilet, treatment of suspected cases and even regular cleaning and disinfection of hands to eliminate bacteria picked up by exchanging greetings, touching money and even contact with tables in classrooms and the working surfaces of offices are a few of the many methods used to prevent and control infections by pathogens transmitted by this route. *S. aureus* is transmitted by this route and successful transmission can lead to skin infections like abscesses, scalded skin syndrome and even food borne gastroenteritis. According to Kimel (1996), scrubbing the body or hands particularly with soaps is the first defense against bacteria and other pathogens that cause such infections hence these methods can be implemented in the control and prevention of *S. aureus* infections (Kimel, 1996).

Antimicrobial agents found in some medicated soaps have the ability to destroy some pathogens like most Staphylococci and methicillin resistant *S. aureus* (MRSA). These agents include: Triclosan, chloroxylenol and Trichlorocarbanilide. Chemotherapy for *S. aureus* is becoming increasingly difficult. Some antibiotics like ampicillin, erythromycin and tetracycline were used for treatment of Staphylococcal infections but with the emergence of multi-drug resistant strains, the use of these drugs for treatment diminishes. The pharmaceutical industry is responding either by modifying existing compounds to broaden their spectra or by developing novel compounds. The result of this response includes oxazolidinones and a combination drug consisting of semisynthetic derivatives of streptogramin A (dalfropristin) with streptogamin B (quinipristin). Vancomycin is often the only effective agent available for therapy especially for the treatment of MRSA infections, but its extensive use may help to promote colonization and infection with

Vancomycin-resistant enterococci. Attempts to find other unique compounds that may attack novel or new bacterial targets are also on the way (Archer, 1998; Anstead *et al*, 2013).

1.2.6 Medical relevance of medicated (antibacterial) soap.

The medical importance of medicated soaps depends on their ability to either kill or inhibit the growth of microorganisms. Some medicated soaps have a broad spectrum of antimicrobial activity on all types of microorganisms (bacteria, viruses, and fungi etc.) while other medicated soaps have activity against a limited group of microorganisms (narrow spectrum of activity). For a soap to be effectively used for the medical purposes, it should be able to ensure protection against a wide range of microorganisms. Soaps are water-soluble or insoluble cleansing agents made from animal and vegetable fats, oils and greases; or chemically, the sodium or potassium salt of a fatty acid formed by the saponification interaction of fats and oils with alkali. Based on the formulation and functional chemical constituents, soaps could be categorized as toilet soaps, antiseptic soaps or medicated soaps. The antiseptic and medicated soaps are made to fight pathogenic microbes and other germs due to special chemical additives in them while the toilet ones are made for conventional cleaning purposes (Ogunnowo *et al*, 2010).

Antibacterial soaps can remove 65 to 85% of bacteria from the human skin with Triclosan, Trichlorocarbanilide, and P-Chloro-in-xylenol (PCMX/Xylenol) being commonly used antibacterials in medicated soaps which elicit this effect (Osborne and Grube, 1982; Larson *et al*, 1989).

Although many people consider that an antimicrobial portion of soaps is effective at preventing communicable disease, recent studies have proven that too much of it can have an opposite effect spreading disease/infection instead of preventing them (Poole, 2002). According to Bettley (1960), some antibacterial soaps can cause nummular eczema, eczematous dermatitis, and eczematous eruption (Bettley, 1960).

1.3 STATEMENT OF THE PROBLEM

Bacterial infection is a public health problem (Johnson *et al*, 2002). The best method for killing or inhibiting their growth is of great importance to man; as such confirmation of the efficient antibacterial activity of Dettol, Pharmapur and Tetmosol which are commonly used medicated soaps is a milestone to achieving this goal.

1.4 HYPOTHESIS

Medicated soaps have different antibacterial strengths.

1.5 OBJECTIVES

1.5.1 General objectives

> To investigate the antibacterial effect of three different brands or types of medicated soaps (Dettol, Pharmapur and Tetmososol) on *Staphylococcus aureus*.

1.5.2 Specific objectives

> To determine the prevalence of *Staphylococcus aureus* from the skin.

> To determine the antimicrobial susceptibility pattern of different concentrations of soap on *Staphylococcus aureus*.

1.6 LIMITATIONS OF STUDY

This study was limited by a number of factors; these included:

- ➤ Absence of an oven for sterilizing and drying of the filter paper discs before inoculation following storage. To reduce this, discs were dried in sterile (Autoclaved), well labelled glass petri dishes.
- ➤ The working laboratory was available for short periods of time. Autoclaved filter papers which were supposed to be dipped in soap solution for an hour to ensure complete saturation of discs with soaps solutions. This was mitigated by dipping the filter papers for 25mins.
- ➤ Absence of a Bunsen burner (flame) for the practice of aseptic techniques. This was mitigated by using an Alcohol lamp.
- ➤ The study was limited to *S. aureus* which is just one of the many pathogens of concern to Public health.

CHAPTER TWO

2.0   **MATERIALS AND METHODS**

This study was carried in the Life Science Laboratory of the Faculty of Science, University of Buea, Cameroon.

2.1 MATERIALS

2.1.1 Materials and instruments

Mannitol salt agar (MSA), Muller Hinton agar (MHA), Medicated soap solutions (Dettol, Pharmapur, Tetmosol), Petri dishes, 70% alcohol, 95% alcohol Gloves, cotton, Aluminium foil, Swab sticks, Test tubes, Alcohol lamp, McCartney bottles, Forceps, Graduated Syringe, Micropipette, pipette tips, electronic balance, 0.5 McFarland standard, Saline (0.85% NaCl), beaker, conical flask, inoculating loop, spreader, ruler, sterile blade, filter papers, medicated soap impregnated discs, Antibiotic discs, Safranin, Crystal violet, iodine, H_2O_2, plasma, distilled water.

2.1.2 Preparation of 70% alcohol

70% alcohol was made by diluting 95% alcohol with water.

Volume (V_1) of 95% alcohol to be diluted= 80mL

Volume (V_2) of water needed=?

Concentration= [from 95% (C_1) to 70% (C_2)]

From $C_1V_1=C_2V_2$;

$V_2= C_1V1/C_2$

$V_2 = 95*80/70$

$= 108.57\text{mL}$

Therefore, for 108.57-80=28.57 hence, 28.57mL of water was added to 80mL of 95% alcohol to give 70% alcohol.

2.1.3 Preparation of soap solutions

One hundred millilitres of water was autoclaved in a bottle and allowed to cool to room temperature. Using an electronic balance and a sterile blade, 1g of soap was scraped from the tablet of each soap (Dettol, Pharmapur, Tetmosol). Using a sterile graduated syringe, 9mL each of sterile water was put in 9 McCartney bottles (3 for the 1st dilution, the next 3 for the 2nd dilution and the last 3 for the 3rd dilution). One gram of each soap was then put in each of the 3 bottles for the 1st dilution, capped with the lids and then labelled. After it had dissolved, serial dilution was then done for the 2nd and 3rd dilution by transferring 1mL soap solution; resulting in different concentrations of each soap solution ranging from 111.1mg/ml to 1.37mg/ml (Obi, 2014).

2.1.4 Preparation of soap impregnated discs

Using a disc borer, a filter paper was bored into 8mm discs and transferred into a closed bottle, then autoclaved. Using sterile forceps, the discs were then immersed into the soap solutions for 25mins after which they were removed and placed in well labelled sterile glass petri dishes and allowed to dry at room temperature.

2.1.5 Preparation of McFarland standard

McFarland standard (0.5) was prepared by using 1% w/v barium chloride ($BaCl_2$) and 1.175% v/v sulphuric acid (H_2SO_4). 0.5mL of 1% w/v $BaCl_2$ was added to 99.5mL of 1.175% v/v H_2SO_4 and mixed.

2.1.6 Media preparation

Assuming that each petri dish should contain 20mL of medium, the media were prepared according to the manufacturer's instructions. 12 plates of Mannitol salt agar and 3 plates of Muller Hinton agar were needed for the study.

2.1.6.1 Preparation of Mannitol salt agar (MSA)

111g → 1000mL

Xg → 240mL

Xg=111g*240mL /1000mL

=26.64g

Mannitol salt agar (26.64g) was measured using an electronic balance and placed in 240mL of water in a beaker and then heated to dissolve completely. The beaker was then covered and the medium was autoclaved at 121°C for 15mins. It was then removed from the autoclave and allowed to cool after which approximately equal amounts were aseptically poured into the 12 petri dishes and allowed at room temperature to solidify. The media-containing petri dishes were then stored in the refrigerator at 4°C until used.

2.1.6.2 Preparation of Muller Hinton agar

$38g \rightarrow 1000mL$

$Xg \rightarrow 60mL$

$Xg = 38g*60mL/1000mL$

$= 2.28g$

Muller Hinton agar (2.28g) was measured using an electronic balance and placed in 60mL of water in a beaker and then heated to dissolve completely. The beaker was then covered and the medium was autoclaved at 121°C for 15mins. It was then removed from the autoclave and allowed to cool after which equal amounts were aseptically poured into the 3 petri dishes and allowed at room temperature to solidify. The media-containing petri dishes were then stored in the refrigerator at 4°C until used.

2.2 METHODS

2.2.1 Sample collection and isolation of microorganisms

The medicated soap samples were purchased from a pharmacy in Buea. Using sterile swab sticks, samples were collected from the anterior nares of the nose, hand and the armpit of 2 human subjects. Six Mannitol salt agar plates were removed from the refrigerator allowed to cool to room temperature. The swab sticks containing the samples were then streaked (streak plate technique) on the plates and labelled. The plates were then incubated at 37°C for 24hrs. After

24hrs suspected colonies were sub-cultured on the remaining 6 mannitol salt agar plates to obtain pure cultures.

2.2.2 Identification and confirmation of isolates

This was done for each of the 6 samples by Gram staining followed by examination under the microscope (oil immersion). Further confirmation was by biochemical testing using catalase and coagulase (Cheesbrough, 2005).

Gram staining: A colony was picked with a sterile (flamed) inoculating loop and emulsified on a slide to make a smear. It was then air dried and heat fixed by passing over the flame. The slide was then floated with 5 drops of crystal violet and allowed for 1min and then washed off. Five drops of iodine was placed on the slide and allowed for 1min and then washed. Next it was decolorized by tilting in alcohol for 5secs. After these, 5 drops of safranin were placed on the slide and then washed off gently. The slide was allowed to dry and examined under the microscope (x1000).

Catalase test: Catalase test was done by putting 2ml of H_2O_2 in a test tube then, a colony was placed in it to observe for effervescence.

Coagulase test: Coagulase test was done by collecting 5ml of venous blood and placed in an EDTA tube and then centrifuged to obtain plasma. A drop of distilled water was dropped on a slide using a dropper and using a flamed wire loop, a colony of the organism that has been checked for positive results of Gram stain was emulsified to make a thick suspension. A loopful of plasma was then added to the suspension and mixed gently (Cheesbrough, 2005).

2.2.3 Antimicrobial susceptibility testing

This was done using the Kirby-Bauer disc diffusion technique (Cheesbrough, 2005).The working bench was sterilized with 70% alcohol. Three Muller Hinton agar (MHA) plates that were stored in the refrigerator were removed and placed on the bench for their temperature to adjust to room temperature. Eight millimetre discs were placed in a little amount of saline and allowed for 25mins (This served as the negative control discs).

Four millilitres of saline solution was measured accurately and transferred into a sterile McCartney bottle. An inoculating loop was then flamed and when it had cooled was used to pick colonies of then test organism (*Staphylococcus aureus*) and transferred into the bottle containing saline solution. This mixture was then adjusted to the turbidity of McFarland standard. This produced a bacterial suspension of approximately same turbidity with that of 0.5 McFarland standard. The 3 plates of MHA were then labelled 10^{-1}, 10^{-2} and 10^{-3} for the 1st, 2nd and 3rd dilutions respectively. Using a micropipette, 100µL of the bacterial suspension was then transferred into each of the agar plates aseptically and a sterile spreader was used to spread suspension. The plates were inverted and allowed for 5mins. Using sterile forceps, the discs impregnated with different medicated soaps were then placed at different positions in the appropriate plate (discs were placed in the plate labelled with the concentration of soap found in them) 25mm apart from each other and 15mm away from the petri dish wall. The antibiotic control discs (Ciprofloxacin) that severed as positive control discs was placed at the center of the 3 different discs and the saline impregnated discs were placed in each plate. The plates were then incubated in inverted positions at 37°C for 24hrs. Results were then obtained in millimetres (mm). Statistical analysis was then done using Analysis of variance (ANOVA).

CHAPTER THREE

3.0   **RESULTS**

Table 1: Morphological and biochemical characteristics of test organism

Colonial description	Gram reaction	Microscopy	Mannitol fermentation	Catalase test	Coagulase test	Possible isolate
Circular, shiny, yellow colonies	+	Gram + grape-like cocci in clusters	+	+	+	*Staphylococcus aureus*

Where + = positive

3.1 Prevalence of *Staphylococcus aureus* on skin of subjects

Of the 6 samples that were investigated, only 4 samples showed growth of *S. aureus* giving a prevalence of 66.67%. 50% of the positive isolates are those from the nose of both subjects, 25% from the armpit of subject 2 and the remaining 25% from the hand of subject 1. This is depicted in Table 2 below.

Table 2: *S. aureus* distribution on the different parts of the skin

Subject	Specimen	Growth results
1	Nose	+
	Armpit	-
	Hand	+
2	Nose	+
	Armpit	+
	Hand	-

Where + = positive and - = negative

3.2 Antimicrobial patterns

This investigation produced results that revealed that all investigated soaps showed antibacterial activity though to different extents. Dettol was found to be the most effective against *S. aureus*, having the highest zone of inhibition (26mm and 8mm at both the highest and lowest dilutions respectively). Tetmosol showed the least antibacterial activity having 13mm at the highest dilution and 6mm at the lowest dilution. This is depicted in table 3.

Table 3: Zone diameter of inhibition (mm) on *S. aureus* by soap samples

Concentration (mg/mL)	Dettol	Pharmapur	Tetmosol
111.1	26	16	13
12.36	12	10	9
1.37	8	7	6

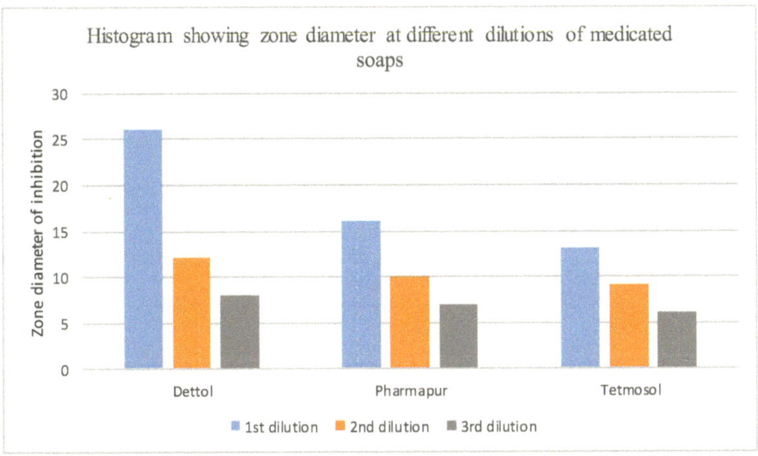

Figure1: Histogram relating the zone diameters at different dilutions of medicated soaps

The negative control sample (saline) showed no observable inhibition against *S. aureus*. This correlates with the absence of active ingredients in its formulation. The positive control (Ciprofloxacin) on the other hand showed considerable antibacterial activity and justifies why it is used for the treatment of many bacterial infections. This is shown in table 4.

Table 4: Diameter of zone of inhibition for positive control (antibiotic disc) and negative control (physiological saline solution)

Code/concentration	Agent	S/mm	I/mm	R/mm	Recording	Conclusion
CIP 30ug	Ciprofloxacin	≥ 21	16-20	≤ 15	40mm	Susceptible
S	Saline				0mm	Resistant

Where;

S= Susceptible, I= Intermediate, R= Resistant (BD BBL Sensi-disc, 2011; Cheesbrough, 2005)

Analysis of variance (ANOVA) for the antibacterial activities (zone diameter) of the medicated soaps at $P<0.05$ showed no statistical significance. This is shown in table 5.

Table 5: ANOVA table verifying significance for zone diameter of inhibition

Differences	Sum of square	Degree of freedom	Mean sum of square	F-statistics	F-critical	Significance
Between groups	57.556	2	28.778			
Within groups	245.333	6	40.889	0.704	5.143	No significance
Total	302.889	8				

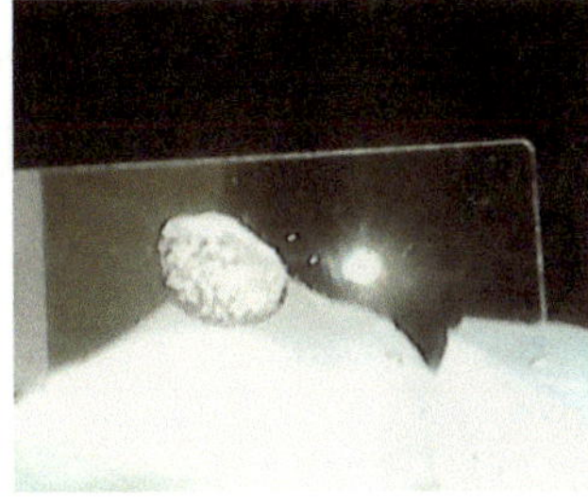

Figure 2: Coagulase positive slide test showing coagulation, for the identification of *Staphylococcus aureus*.

Figure 3: 0.5 McFarland standard used for adjusting the turbidity of bacterial suspension.

Figure 4: Catalase test positive result for the identification of Staphylococci.

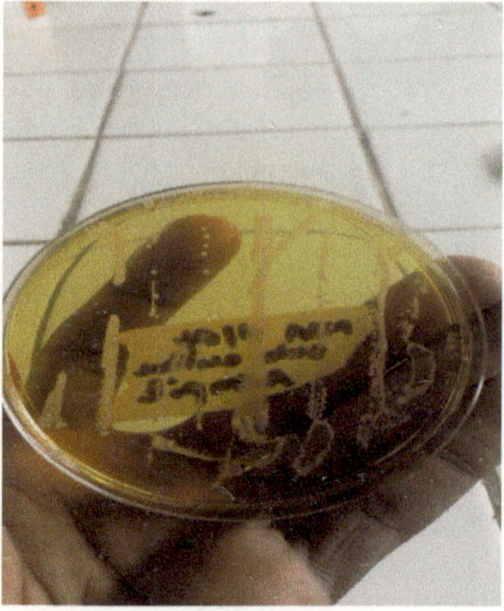

Figure 5: Yellow, circular colonies fermenting mannitol by *Staphylococcus aureus* on mannitol salt agar (streak plate technique).

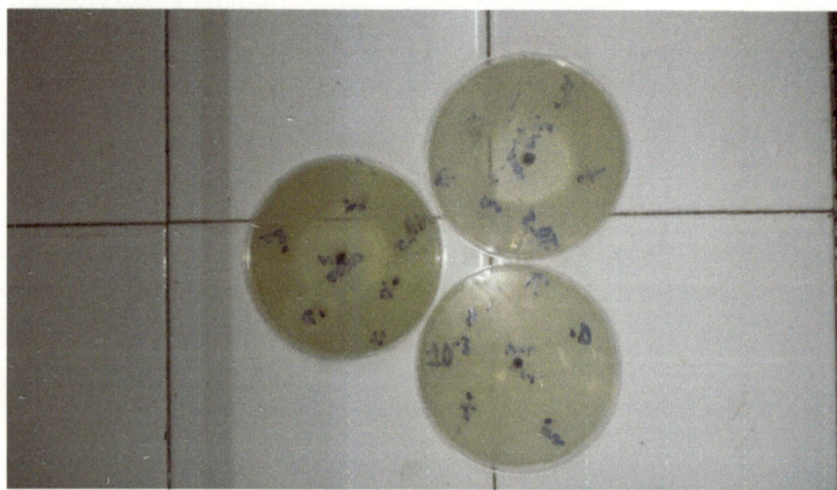

Figure 6: Muller Hinton agar plates in which Antimicrobial susceptibility testing was done.

CHAPTER FOUR

4.0   **DISCUSSIONS**

Washing with soap should not only be a means of self-protection against bacterial infections but should also be an obligation that humans owe others especially the health care personnel. According to Riaz *et al*, (2009) washing of hands prior to surgery and examination of patients in order to remove some potentially harmful transient flora as well as to reduce a number of harmful flora which may cause opportunistic infections should be routinely practiced. Soaps that contain antimicrobial agents can be a powerful tool for this practice.

When people purchase medicated soaps, they are faced with the problem of choice with many ending up asking questions to owners of cosmetic shops who have little or no knowledge about the antimicrobial activity of the soaps they sell. This study attempts to guide individuals on which soap is better for use. Although the indicated soaps in this study were found to possess different antibacterial strength, they all possess chemical agents/compounds that can effectively remove or kill bacteria thus, preventing infection.

Triclosan and Trichlocarban which are ingredients found in the soaps used in this study are popular active ingredients used in medicated soaps. These chemical compounds function by denaturing cell activity and interfering with microbial metabolism but their activity depends on a number of factors such as: inherent properties of the organism, contact time, concentration, individual formulation and skin sensitivity (Obi, 2014). However, according to this study, Dettol showed the greatest overall antistaphylococcal activity followed by Pharmapur then Tetmosol (Dettol>Pharmapur>Tetmosol). Triclosan acts not only as an antibacterial agent but also as an antiviral and antifungal agent which is bacteriostatic at low concentrations and bactericidal at high concentrations and has particular activity in inhibiting growth of gram positive bacteria (Aiello, 2007; Obi CN, 2014). Of the 3 soaps, Triclosan is found only in Dettol, Trichlocarban is found in both Dettol and Pharmapur and neither Triclosan nor Trichlocarban is found in Tetmosol which is a possible explanation for the trend in antibacterial activity observed (Dettol>Pharmapur>Tetmosol) in this study. The best antibacterial activity of Dettol can be attributed to the unique Triclosan formulation of Dettol.

4.1 Conclusion

- *S. aureus* is widely distributed transiently on different areas of the human skin.
- The medicated soaps investigated in this study showed different patterns of activity against *S. aureus*. Dettol among the others can be used to prevent nosocomial and community-acquired *S. aureus* skin infections.
- Lack of statistically significant difference between the zones of inhibition maybe due to the small sample size.

4.2 Recommendation

Aside from Vancomycin that has been used for treatment of *S. aureus* infections, Ciprofloxacin can also be used as a good chemotherapeutic agent. Compound in these soaps act by denaturing cell membrane components and interfering with microbial metabolism, hence the irrational and long term use of these soaps should be discouraged so as to avoid resistance. The minimum inhibitory concentration (MIC) and minimum bactericidal concentration (MBC) of the soap samples investigated was not done, hence much needs to be done to determine these concentrations.

REFERENCES

Aiello AE, Larson EL and Levy SB (2007). Consumer antibacterial soaps: Effective or just risky. *Clinical Infectious Diseases*. 45: 137-147.

Aly R and Maibach HI (1976). Effect of antimicrobial soap containing Chlorhexidine on the microbial flora of skin. *Applied and Environmental Microbiology*. 31: 931-935.

Anstead GM, Cadena J and Javeri H (2013). Treatment of infections due to resistant *Stapylococcus aureus*, Methicillin-Resistant *Staphylococcus aureus* (MRSA) Protocols. *Methods in Molecular Biology*. 1085: 259-309

Archer GL (1998). *Staphylococcus aureus*: A Well-Armed Pathogen. *Clinical Infectious Diseases*. 26:1179-1181.

BD BBL Sensi-disc (2011). Antimicrobial susceptibility test discs. 1-16.

Bettley FR (1960). Some effects of soap on skin. *British Medical Journal*. 1675-1679.

Bhat RP, Prajna, Menezez VP and Shetty P (2011). Antimicrobial activities of soaps and detergents. *Advances in Bioresearch*. 2(2): 52-62.

Cheesbrough M (2005). District Laboratory Practice in Tropical countries. *Cambridge University Press, Cambridge*. 2: 1-434.

Eiff CV, Becker K, Machka K, Stammer H, and Peters G (2001). Nasal carriage as a source of *Staphylococcus aureus* bacteraemia. *New England Journal of Medicine*. 344(1):11-16.

Grice EA and Serge JA (2011). The skin microbiome. *Nature Reviews Microbiology*. 9(4): 244-253.

Harris LG, Foster SJ and Richards RG (2002). An introduction to *Staphylococcus aureus* and techniques for identifying and quantifying *S. aureus* adhesins in relation to adhesion to biomaterials: review. *European Cells and Materials*. (4): 39-60.

Johnson SA, Goddard PA, Ilife C, Timmens B, Richard AH, Robson G and Handley PS (2002). Comparative susceptibility of resident and transient hand bacteria to Parachlorometa-xylenol and triclosan. *Journal of Applied Microbiology*. 93:336-344.

Kimel LS (1996). Hand washing education can decrease illness absenteeism. *Journal of School Nursery*. 12: 14-18.

Larson E (1999). Skin hygience and infection prevention: More of same or different approaches. *Clinical Infectious Diseases*. 29: 1287-1294.

Larson E, McGinley K, Grove GL, Leydenn JJ and Talbot GH (1989). Physiology, microbiologic, and seasonal effects of handwashing on the skin of health care personnel. *American Journal of Infection Control*. 14: 5-90.

Lowy FD (1998). *Staphylococcus aureus* infections. *New England Journal of Medicine*. 339(8): 520-532.

Lyon BR and Skurray R; (1987). Antimicrobial resistance *of Staphylococcus aureus*: Genetic basis. *Microbiological Reviews*. 51(1): 88-134.

Murray PR, Baron EJ, Jorgensen JH, Pfaller MA and Yolken RH (2003). *Manual of Clinical Microbiology*, 8th edn. Washington,DC: American Society for Microbiology.

Murray PR, Lim T, Pearson J, Grubb W and Lum G (2004). Community-onset of methicillin-resistant *Staphylococcus aureus* bacteraemia in Northern Australia. *International Journal of Infectious Dis*eases. 8(5):275-283.

Naber CK (2009). *Staphylococcus aureus* Bacteraemia: Epidemiology, Pathophysiology and Management Strategies. *Clinical Infectious Diseases*. 48(4): 231-237.

Nkwelang G, Akoachere J-F TK, Kamga LH, Nfoncham ED, and Ndip RN (2009). *Staphylococcus aureus* isolates from clinical and environmental samples in a semi-rural area of Cameroon: phenotypic characterization of isolates. *African Journal of Microbiology Research*. 3(11): 731-736.

Noble WC and Somerville DA (1974). Microbiology of human skin. W.B. Saunders, Philadephia. 50-76.

Obi CN (2014). Antibacterial activity of some medicated soaps on selected human pathogens. *American Journal of Microbiological Research*. 2(6): 178-181.

Ogunnowo AA, Esekiel CN, Anokwuru CP, Ogunshola YA and Kuloyo OO (2010). Anticandidal and antistaphylococcal activity of soap fortified with *Ocimum gratissimum* extract. *Journal of Applied Sciences*. 10(18): 2121-2126.

Osborne RC and Grube J (1982). Hand disinfection in dental practice. *Journal of Clinical Preventive Dentistry*. 4: 11-15.

Poole k (2002). Mechanisms of bacterial bioade and antibiotic resistance. *Journal of Applied Microbiology.* 92: 555-564.

Riaz S, Admad A and Hasnain S (2009). Antibacterial activity of soaps against daily encountered bacteria. *African Journal of Biotechnology.* 8(8): 1431-1436.

Rotter ML (1966). Handwashing and hand disinfection. *Hospital Epidemiology and Infection Control.* 1052-1068.

Shorr AF and Lodise T (2006). Burden of methicillin-resistant *Staphylococcus aureus* on healthcare cost and resource utilization. *ISMR.* 1: 4-11.

APPENDIX

APPENDIX I: Reagents and Culture Media Formulation and Preparation

REAGENTS

Crystal violet

Formulation

Crystal violet (dye)................... 20g

Ammonium oxalate.................... 9g

Ethanol................................95mL

Distilled water........................ 1L

Twenty grams of the dye was dissolved in 95millilitres of ethanol and 9grams of ammonium oxalate was dissolved in a liter of distilled water. They were mixed then filtered.

Lugol's iodine

Formulation

Potassium iodide.............. 20g

Iodine...........................10g

Distilled water................. 1L

Twenty grams of potassium iodide was dissolved in a liter of distilled water and 10grams of iodine added to the solution and mixed.

Safranin

Formulation

Safranin.................0.25mL

Ethanol (96%)......... 10mL

Safranin (0.25g) was dissolved in 10mL of 95% alcohol solution.

Hydrogen peroxide (3%)

Hydrogen peroxide is prepared by the electrolysis of ammonium hydrogen sulphate dissolved in excess of sulphuric acid, using platinum electrolyte and a high current density. In order to have high current density (that is, current strength per unit area of electrode) at the anode, the area of the anode should be small. In this process ammonium per sulphate is produced by oxidation at the anode while hydrogen is liberated at the cathode. The solution containing ammonium per sulphate is heated at 43mm pressure when it hydrolysis, yielding hydrogen peroxide.

$$(NH_4)_2S_2O_8 + 2H_2O = 2NH_4HSO_4 + H_2O_2$$

Hydrogen peroxide along with water distills over. The aqueous solution ($3\%H_2O_2$) is formed.

CULTURE MEDIA FORMULATION

Mannitol salt agar

Ingredient and Composition

Pancreatic digest of casein………..5.0g

Peptic digest of animal tissue……...5.0g

Beef extract……………………... .10.0g

Sodium chloride…………………...75.0g

D-mannitol…………………….…..10.0g

Phenol red………………….………25.0mg

Agar………………………………….15.0g

Muller Hinton agar

Ingredients and composition

Beef extract...............................2.0g

Acid hydrolysate of casein................17.5g

Starch..1.5g

Agar..17.0g

Distilled water............................1000mL

APPENDIX II: Soap ingredient and composition

Dettol original medicated soap

Ingredients and composition

Sodium Palmate, Sodium Palm Kernelate, Talc, Aqua, Pinus Palustris Oil, Sodium C14-16 Olefin Sulfonate, Triclosan, Chloroxylenol, Triclocarban, Glycerin, Tetrasodium EDTA, Tetrasodium Etidronate, Limonene, PEG-7 Amodimethicone, Trideceth-10, Sodium Laureth Sulfate, Hydroxypropyl Cyclodextrin, Tetrabutyl Ammonium Bromide, Acetic Acid, Sodium o-Phenylphenate, Methylchloroisothiazolinone, Cl 77891, Cl 11680.

Pharmapur Protex medicated soap

Ingredients and composition

Sodium Palmate *A, Aqua, Sodium Tallowate *B, Sodium Palm Kernelate, Alternate/Talc, Glycerin, PEG-12, Stearic Acid, Parfum, Sodium Chloride, Triclocarban, Pentaerithrityl Tetra-di-t-butyl Hydroxyhydrocinnamate, Pentasodium Pentetate, Thymus Vulgaris Extract, Calendula

Officinalis Extract, Camelis Sinensis Leaf Extract, Rosmarinus Officinalis Extract, Melalueca Alternifolia Oil, Alpha-Isomethyl Ionone, Benzyl Salicylate, Butylphenyl Methylpropional, Hexyl Cinnamal, Limonene, Linalool, Cl 77891, Cl 47005, Cl 61570.

Tetmosol medicated soap

Ingredients and composition

Soap Base, Monosulfiram B.P 5%w/w, Citronella Oil

YOUR KNOWLEDGE HAS VALUE

- We will publish your bachelor's and
  master's thesis, essays and papers

- Your own eBook and book -
  sold worldwide in all relevant shops

- Earn money with each sale

Upload your text at www.GRIN.com
and publish for free